2019

室内设计模型集成

简约风格家居

INTERIOR DESIGN MODEL LIBRARY

SIMPLE STYLE · HOME

海峡出版发行集团 | 福建科学技术出版社

作者简介 AUTHOR PROFILE

叶 斌 / Ye Bin

高级建筑师
国广一叶装饰机构首席设计师
福建农林大学兼职教授
南京工业大学建筑系建筑学学士
北京大学 EMBA
中国室内设计学会理事

获奖设计作品

余韵	2018 第十二届中国国际室内设计双年展金奖
無·色	2018 第十二届中国国际室内设计双年展金奖
TIMES	2018 第十二届中国国际室内设计双年展银奖
简·木	2018 第十二届中国国际室内设计双年展银奖
长乐禅修中心	2018 第十二届中国国际室内设计双年展银奖
灵·动	2018 第十二届中国国际室内设计双年展银奖
长乐禅修中心	2016/2017 APDC 亚太室内设计精英邀请赛展览空间方案类大奖
听海	2016/2017 APDC 亚太室内设计精英邀请赛住宅空间工程类大奖
FORUS VISION	2017 年第二十届中国室内设计大奖赛零售商业类 金奖
听海	2017 年第二十届中国室内设计大奖赛住宅类 铜奖
爱家微运动公社	2017 年第七届中国国际空间设计大赛（中国建筑装饰奖）娱乐会所空间方案类 金奖
皇帝洞廊桥主题酒店	2017 年第七届中国国际空间设计大赛（中国建筑装饰奖）酒店空间方案类 银奖
一隅绿意	2017 年第七届中国国际空间设计大赛（中国建筑装饰奖）办公空间方案类 铜奖
世欧澜山	2017 年第七届中国国际空间设计大赛（中国建筑装饰奖）商品房、样板房空间方案类 铜奖
长乐电力大楼	2015~2016 年度中国建筑工程装饰奖（公共建筑装饰类）
叶禅赋	2016 第十一届中国国际室内设计双年展金奖
FORUS	2016 第十一届中国国际室内设计双年展金奖
Lee House	2016 第十一届中国国际室内设计双年展银奖
静·念	2016 第十一届中国国际室内设计双年展银奖
仕林东湖	2016 第十一届中国国际室内设计双年展铜奖
白说》	2016 第十一届中国国际室内设计双年展铜奖
"一扇窗，漫一室"	2016 第十一届中国国际室内设计双年展铜奖
溪山温泉度假酒店（实例）	2014 年第十届中国国际室内设计双年展金奖
正兴养老社区体验中心	2014 年第十届中国国际室内设计双年展银奖
永福设计研发中心	2014 年度全国建筑工程装饰奖（公共建筑装饰设计类）
宇洋中央金座	2013 年第十六届中国室内设计大奖赛铜奖
宁德上东曼哈顿售楼部	2013 年第四届中国国际空间环境艺术设计大赛（筑巢奖）优秀奖
福建洲际酒店	2012 年首届亚太金艺奖酒店设计大赛金奖
瑞莱春堂	2012 年第四届"照明周刊杯"照明应用设计大赛金奖
前线共和广告	2012 年第十五届中国室内设计大奖赛金奖
前线共和广告	2012 年第九届中国室内设计双年展金奖
阳光理想城	2012 年第九届中国室内设计双年展金奖
福州情·聚春园	2012 年第九届中国室内设计双年展银奖
宁化世界客属文化交流中心	2012 年第九届中国室内设计双年展铜奖
映·像	2012 年第二十届亚太室内设计大奖赛铜奖
名城港湾 157#103	2012 年第三届中国国际空间环境艺术设计大赛（筑巢奖）优秀奖
一信（福建）投资	2011 年第十四届中国室内设计大奖赛金奖
福建科大永和医疗机构	2011 年中国最成功设计大赛最成功设计奖
素丽娅泰 SPA	2010 年第八届中国室内设计双年展金奖
摩卡小镇售楼中心	2010 年第八届中国室内设计双年展银奖
大洋鹭洲	2010 年第八届中国室内设计双年展铜奖
素丽娅泰 SPA	2010 年亚太室内设计双年展大奖赛商业空间设计银奖
繁都魅影	2010 年亚太室内设计双年展大奖赛住宅空间设计银奖
繁都魅影	2010 年亚洲室内设计大奖赛铜奖
中央美苑	2010 海峡两岸室内设计大赛金奖
繁都魅影	2010 海峡两岸室内设计大赛金奖
光.盒中盒	2010 海峡两岸室内设计大赛金奖
皇帝洞书库	2009 年"尚高杯"中国室内设计大奖赛二等奖

另 129 项设计作品荣获福建省室内设计大奖赛一等奖、金奖。

荣 誉

当选 2013-2015 年度福建省最具影响力设计师（排名第一）
荣获"中国室内设计杰出成就奖"
当选 2009 "金羊奖"中国十大室内设计师
当选中国建筑装饰行业建国 60 年百名功勋人物
当选 1989~2009 中国杰出室内设计师
当选 1997~2007 中国家装十年最具影响精英领袖
当选 1989~2004 全国百位优秀室内建筑师
当选 2004 年度中国室内设计师十大封面人物
当选 2002 年福建省室内设计十大影响人物（第一席位）

著 作

1. 《室内设计图典》(1、2、3)
2. 《装饰设计空间艺术·家居装饰》(1、2、3)
3. 《装饰设计空间艺术·公共建筑装饰》
4. 《建筑外观细部图典》
5. 《国广一叶室内设计模型库·家居装饰》(1、2、3)
6. 《国广一叶室内设计模型库·公建装饰》
7. 《国广一叶室内设计》
8. 《国广一叶室内设计模型库构成元素》(1、2)
9. 《室内设计立面构造艺术》系列
10. 《国广一叶室内设计模型库》系列
11. 《家居装饰·平面设计概念集成》
12. 《概念家居》、《概念空间》
13. 《2009 室内设计模型》系列（5 册）
14. 《2010 家居空间模型》系列（3 册）
15. 《2010 公共空间模型》系列（2 册）
16. 《2011 家居空间模型》系列（3 册）
17. 《2011 公共空间模型》
18. 《2012 室内设计模型集成》系列（5 册）
19. 《2013 公共空间模型集成》系列（2 册）
20. 《2013 家居空间模型集成》系列（3 册）
21. 《2014 空间模型集成》系列（5 册）
22. 《2015 室内设计模型集成》系列（5 册）
23. 《2015 名家家装新图典》系列（3 册）
24. 《2016 公共空间模型库》
25. 《2016 家居空间模型库》系列（4 册）
26. 《新家居装修与软装设计》系列（4 册）
27. 《2017 公共空间模型库》
28. 《2017 家居空间模型库》系列（4 册）
29. 《经典家居设计》系列（4 册）
30. 《2018 年室内设计模型集成》系列（4 册）

叶 猛 / Ye Meng

国广一叶装饰机构副总设计师
国家一级注册建筑师
国家一级注册建造师
中国建筑学会室内分会会员
福建工程学院建筑与规划系讲师
福州大学建筑系学士
中南大学土建学院建筑学硕士

获奖设计作品

融信澜郡	2017 第八届中国国际空间环境艺术设计大赛（筑巢奖）优秀奖
仕林东湖	2016 第十一届中国国际室内设计双年展银奖
融信大卫城—禅韵	2016 福建省室内设计大赛居室空间类金奖
东方韵	2015 中南地区国际空间环境艺术设计大赛方案设计空间铜奖
雅韵·世欧澜山	2015 中南地区国际空间环境艺术设计大赛住宅空间优秀奖
风尚	2015 年度国际空间设计大奖·艾特奖 最佳公寓设计入围奖
融信大卫城	2014 年第十届中国国际室内设计双年展优秀奖
三盛国际公园	2014 年第五届中国国际空间环境艺术设计大赛（筑巢奖）提名奖
名城港湾	2014 年第五届中国国际空间环境艺术设计大赛（筑巢奖）优秀创意奖
融侨外滩	2014 年第五届中国国际空间环境艺术设计大赛（筑巢奖）优秀创意奖
鳌峰洲小区—19A	2013 年第四届中国国际空间环境艺术设计大赛（筑巢奖）优秀奖
阳光理想城	2012 年第九届中国国际室内设计双年展金奖
大洋鹭洲	2010 年第八届中国室内设计双年展铜奖
繁都魅影	2010 年亚洲室内设计大奖赛铜奖
福建工程学院建筑系新馆	2009 年中国室内空间环境艺术设计大赛一等奖
福建工程学院建筑系新馆	2009 年福建室内与环境设计大奖赛公建工程类最高奖
文化主题酒店	2008 年福建省第六届室内与环境设计大赛一等奖
点房财富中心	2007 年"华耐杯"中国室内设计大奖赛二等奖
大家会馆（实例）	2006 年第六届中国室内设计双年展金奖
金钻世家某单元房	2006 年第六届中国室内设计双年展银奖

另出版《建筑外观细部图典》《室内设计图像模型》等著作数十种

前言 PREFACE

国广一叶装饰机构作为"全国最具影响力室内设计机构"（中国建筑学会室内设计分会颁发）、2017年第二十届中国室内设计大奖赛"最佳设计企业"（中国建筑学会室内设计分会颁发）、"2016年度中国建筑装饰杰出住宅空间设计机构"（中国建筑装饰协会颁发）、"2015年度中国建筑装饰设计机构50强企业"（中国建筑装饰协会颁发）、"2013住宅装饰装修行业最佳设计机构"（中国建筑装饰协会颁发）、"2013年度全国住宅装饰装修行业百强企业"（中国建筑装饰协会颁发）、"2012～2013年度全国室内装饰优秀设计机构"（中国室内装饰协会颁发）、"2012年中国十大品牌酒店设计机构"（中外酒店论证颁发）、"2013中国住宅装饰装修行业最佳设计机构"（中国建筑装饰协会颁发）、"1989～2009年全国十大室内设计企业"（中国建筑协会室内设计分会颁发）、"1988～2008年中国室内设计十佳设计机构"（中国室内装饰协会颁发）、"1997～2007年中国十大家装企业"（中国建筑装饰协会颁发）、"福建省建筑装饰装修行业龙头企业"（福建省人民政府闽政文〔2014〕26号宣布）、"福建省建筑装饰行业协会会长单位"，荣获国际、国家及省市级设计大奖上千项。

国广一叶装饰机构首席设计师叶斌荣获"中国室内设计杰出成就奖"、两次荣获"中国十大室内设计师"称号；叶猛被评为"1989～2009年中国优秀设计师"、"福建十大杰出（住宅空间）设计师"称号；另外，51名设计师被评为中国装饰设计行业优秀设计师，126名设计师分别被评为福建省优秀设计师、福州市优秀设计师，120名在职设计师分别荣获历届全国、福建省、福州市室内设计一等奖……

以上荣誉的获得与国广一叶装饰机构23年的设计从业经验有关，国广一叶装饰机构拥有近300人的优秀设计师团队，设计师们通过效果图将优秀设计创意淋漓尽致地表现出来。

自2004年至今，国广一叶装饰机构在福建科学技术出版社已陆续出版了20套模型系列图书，共58本模型系列图书，一直受到广大读者的支持与厚爱。为了不辜负广大读者的期望，我们继续推出《2019室内设计模型集成》系列图书。这系列图书汇集了国广一叶装饰机构2018年制作的1200多个风格各异的室内设计效果图及其对应的3ds Max场景模型文件，可用于读者做室内设计时的有益参考。

本书配套光盘的内容包含效果图原始3ds Max模型和使用到的所有贴图文件。由于3ds Max软件不断升级，此次的模型我们采用3ds Max2011版本制作。模型按图片顺序编排，易于查阅和调用。只有能对模型进一步调整才能体现其价值和生命力，因此提供的3ds Max模型是真正有价值、可随时提取调整用的部分。必须说明的是，书中收录的效果图均为原始模型经过Lightscape渲染和Photoshop后期处理过的成图，是为读者了解后处理效果提供直观准确的参考，与3ds Max直接渲染的效果有一定区别。

著 者
2018年12月

As a well-known decoration company, Guoguangyiye Decoration Group have acquired thousands of international, national and provincial design awards, such as "Top 50 architectural decoration company in China(2015, honored by China Building Decoration Association, CBDA)", "the best design institutions of residential ornament industry in 2013 (honored by China Building Decoration Association, CBDA)", "the Top 100 enterprises of Chinese residential ornament industry in 2013(honored by China Building Decoration Association, CBDA)" "Outstanding Interior Design Companies in China(2012-2013, honored by China National Interior Decoration Association, CIDA)", "Top 10 Candlewood Design Companies in China(2012, honor by Chinese and Foreign Hotel Argument)", "The Best Interior Decoration Association of Chinese Home Decoration(2013, honored by CBDA)", "Top 10 Interior Design Companies in China (1989~2009)", "Top 10 China Interior Design Institutions (1988~2008)", "2012 China top 10 Hotel Design Institutions", "China Top 10 Home Decoration Enterprises (1997~2007)", "Well-Known Brand of Fujian", "the leading enterprises of architectural ornament industry in Fujian province (issued by the people's Government of Fujian Province〔2014〕No. 26)" and "the president company of Architectural Ornament Industry Association of Fujian province".

In Guoguangyiye Decoration Group, a dozen of 51 architects have be granted as "Excellent designer in China's decorative design industryNational 19 architect of China", and 83 126 architects have awarded as "Excellent Architect of Fujian province/Fuzhou", 76 120 architects have won top prize of national, Fujian provincial or Fuzhou. The chief architect Mr. Bin Ye has wined the award of "Distinguished Achievement Award of Chinese Interior Design", and awarded twice "China Top 10 Interior Design Architect". Mr. Meng Ye was awarded "Outstanding Architect of China (1989~2009)".

These above honors are benefitted from the 23 years' design experiences of Guoguangyiye Decoration Group. Guoguangyiye Decoration Group has a team of nearly 300 outstanding designers who can show their outstanding design ideas completely through the renderingsNaturally these achievements have been accomplished because of the high level interior designs of Guoguangyiye, but obviously cannot be attained without high level professional effect drawing that presents the design intent of architects incisively and vividly. Therefore as a product of the collective efforts of architect and graphic designer, it is closely related to the success of project design.

Since 2004, Guoguangyiye has published nineteen twenty series of books on design model database with Fujian Science and Technology Press and 58 series of model books which obtain readers' incredible support and affection. In order to live up to the expectations of the readers, we will continue to publish the series of books《2019 Interior Design Model Library》. all of them have gained wide popularity by their richness and practicality. Therefore, this year we will continue to publish 2018 Interior Design Model Library. This new series consists of over 1900 1200 chic 3ds Max scenario models of various style interior designs design renderings created by Guoguangyiye Decoration Group during 2016~20172018~2019. Being a model databaselibrary, they could also be used as beneficial references for interior design.

The enclosed DVD contains original 3ds Max models of decoration effect drawings and all the map files used in order to create them. Due to the continuous upgrading of 3ds Max software, version 2011 was adopted in the drawing of these models which are arranged in the order of the pictures to make them easily lookup and call. Since as only models that can be further adjusted are valuable, the 3ds Max moulds provided are all of true value and readily available. It should be noted that, all the effect drawings in the books are pictures rendered by lightscape and dealt with by Photoshop, to give an intuitive and precise reference for readers on the after effects which are different from those rendered directly by 3ds Max.

December 2018

CONTENTS 目录

005 / 客厅
LIVING ROOM

097 / 卧室
BED ROOM

153 / 其他功能空间
OTHER ROOM

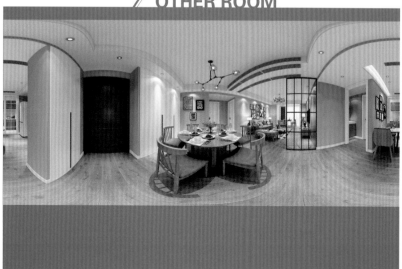

客厅 LIVING ROOM

客厅 | LIVING ROOM | 007

客厅 | 009
LIVING ROOM

客厅 | 011
LIVING ROOM

客厅 | LIVING ROOM | 013

客厅 | LIVING ROOM 015

020

021

客厅 | LIVING ROOM | 017

022

023

024

025

客厅 | LIVING ROOM 019

027

026

028

客厅 | LIVING ROOM | 021

029

032

031

033

客厅 | LIVING ROOM 023

036

035

客厅 | LIVING ROOM | 025

037

039

038

040

客厅 | LIVING ROOM | 027

046

047

客厅 | 029
LIVING ROOM

048

049

客厅 | LIVING ROOM | 031

客厅 | LIVING ROOM | 033

057

客厅 | LIVING ROOM

客厅 | LIVING ROOM | 037

060

061

客厅 | 041
LIVING ROOM

064

067

066

068

客厅 | LIVING ROOM 043

071

072

客厅 | LIVING ROOM | 045

073

075

074

076

077

客厅 | LIVING ROOM | 047

079

078

客厅 | LIVING ROOM | 049

080

082

083

084

客厅 | 051
LIVING ROOM

085

086

087

客厅 | LIVING ROOM | 053

088

090

091

客厅 | LIVING ROOM | 055

092

094

095

096

097

客厅 | LIVING ROOM | 061

102

104

客厅 | LIVING ROOM | 067

114

116

115

118

119

客厅 | LIVING ROOM | 071

124

122

125

123

126

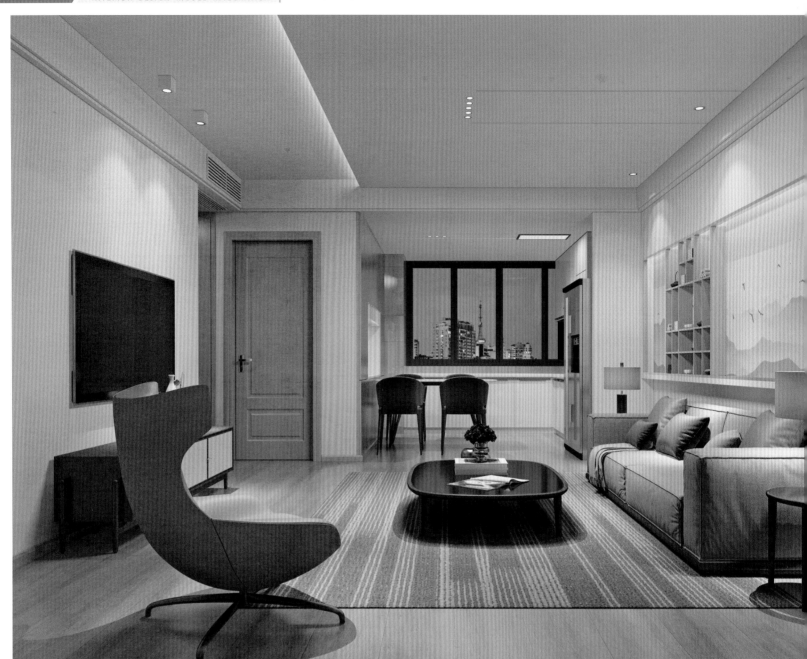

客厅 | LIVING ROOM 073

客厅 | LIVING ROOM 075

132

133

134

客厅 | LIVING ROOM 077

136

135

137

客厅 | LIVING ROOM | 079

138

140

141

142

客厅 | LIVING ROOM | 081

客厅 | LIVING ROOM | 083

145

147

149

客厅 | LIVING ROOM | 087

156

158

157

客厅 | LIVING ROOM | 095

159

161

卧室 BED ROOM

165

166

卧室 | 101
BEDROOM

172

173

175

176

177

178

179

180

卧室 | 107
BEDROOM

181

183

182

184

卧室 | BEDROOM

189

190

193

卧室 | 113
BEDROOM

194

196

195

卧室 | 115
BEDROOM

197

200

199

201

202

204

卧 室
BEDROOM | 117

203

205

206

207

208

卧室 | 119
BEDROOM

209

210

211

214

215

216

218

卧室 | 123
BEDROOM

217

219

220

卧室 | 125
BEDROOM

222

224

225

卧室 | 129
BEDROOM

231

230

232

卧室 | 131
BEDROOM

卧室 | BEDROOM 133

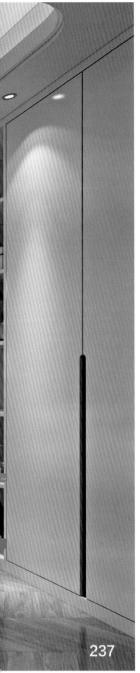

237

240

239

241

244

245

246

247

卧室 | 137
BEDROOM

251

253

卧室 | 139
BEDROOM

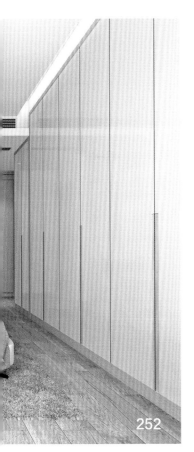

卧室 | 141
BEDROOM

256

259

258

260

卧室 | 143
BEDROOM

261

264

263

265

卧室 | 145
BEDROOM

卧室 | 147
BEDROOM

270

273

272

274

275

276

278

279

卧室 | 151
BEDROOM

280

281

282

283

284

其他功能空间 OTHER ROOM

其他功能空间 | 155
OTHER FUNCTIONAL SPACE

286

287

其他功能空间 | 159
OTHER FUNCTIONAL SPACE

290

292

293

291

294

其他功能空间 | 161
OTHER FUNCTIONAL SPACE

295

298

297

299

300

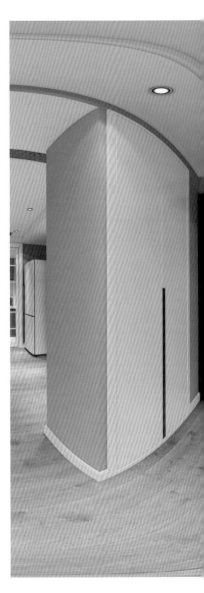

302

其他功能空间 | 163
OTHER FUNCTIONAL SPACE

301

303

304

其他功能空间 | 165
OTHER FUNCTIONAL SPACE

305

306

307

其他功能空间 | **167**
OTHER FUNCTIONAL SPACE

308

311

310

312

其他功能空间 | 169
OTHER FUNCTIONAL SPACE

313

315

316

其他功能空间 | 171
OTHER FUNCTIONAL SPACE

317

319

320

其他功能空间 | 175
OTHER FUNCTIONAL SPACE

图书在版编目（CIP）数据

2019室内设计模型集成．简约风格家居/叶斌，叶猛著．— 福州：福建科学技术出版社，2019.3
 ISBN 978-7-5335-5809-3

Ⅰ.①2… Ⅱ.①叶…②叶… Ⅲ.①住宅-室内装饰设计-图集 Ⅳ.①TU238.2-64

中国版本图书馆CIP数据核字（2019）第018696号

书　　名	2019室内设计模型集成　简约风格家居
著　　者	叶斌　叶猛
出版发行	福建科学技术出版社
社　　址	福州市东水路76号（邮编350001）
网　　址	www.fjstp.com
经　　销	福建新华发行（集团）有限责任公司
印　　刷	恒美印务（广州）有限公司
开　　本	635毫米×965毫米　1/8
印　　张	22
图　　文	176码
版　　次	2019年3月第1版
印　　次	2019年3月第1次印刷
书　　号	ISBN 978-7-5335-5809-3
定　　价	288.00元

书中如有印装质量问题，可直接向本社调换